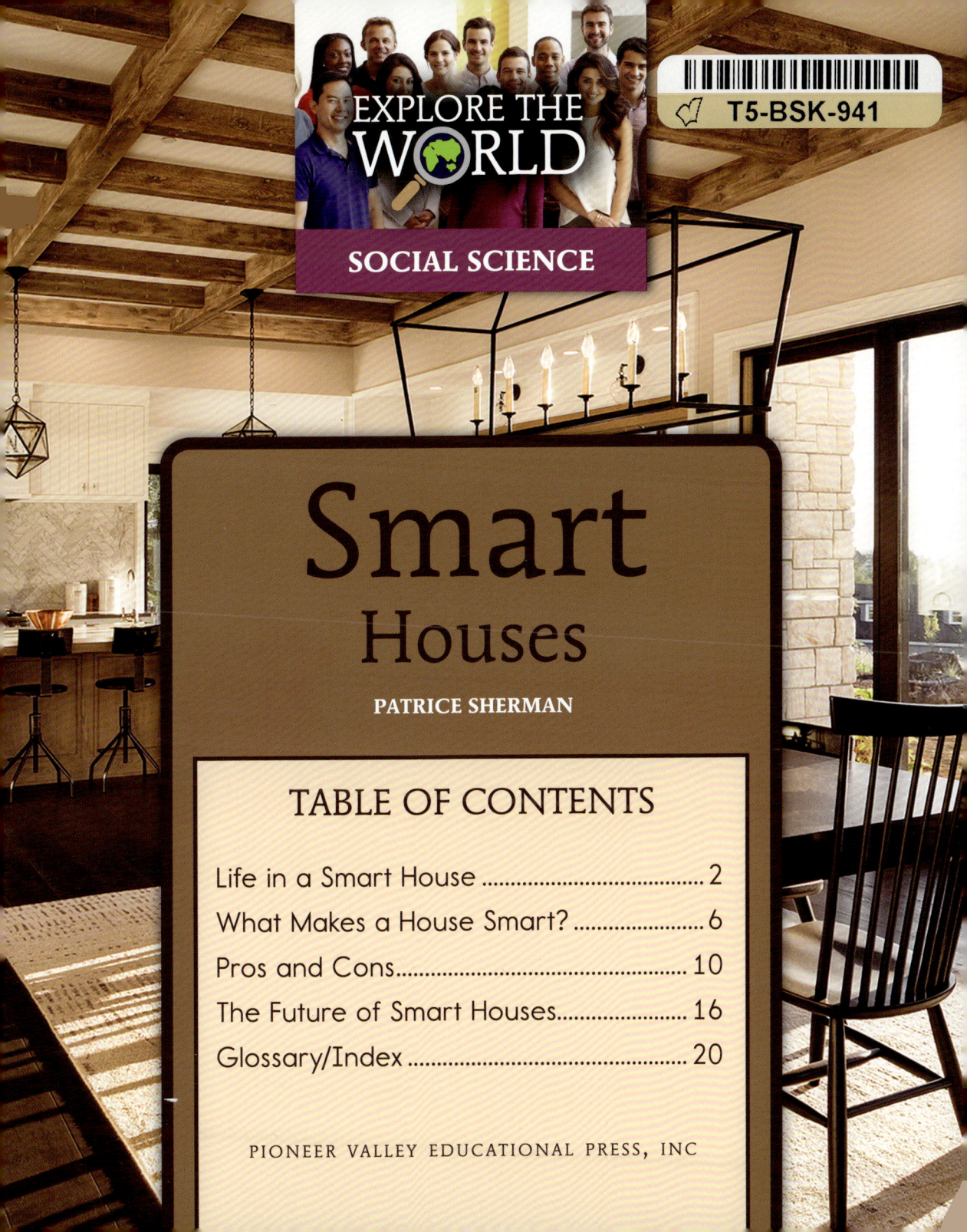

Smart Houses

PATRICE SHERMAN

TABLE OF CONTENTS

Life in a Smart House 2
What Makes a House Smart? 6
Pros and Cons 10
The Future of Smart Houses 16
Glossary/Index 20

PIONEER VALLEY EDUCATIONAL PRESS, INC

LIFE IN A SMART HOUSE

What is it like to live in a smart home? Imagine you are living in the future...

As soon as you wake up in the morning and start moving around, the temperature of your room **automatically** adjusts to your needs. If it's winter, the heat turns on so you can be comfortable while you get dressed. In the summer, the air conditioner may kick in to make it a few degrees cooler. When you step into the shower, you say, "Start," and a gentle spray of water turns on. When you're finished, you say, "Stop."

"Zino, what's the weather today?" you ask your smart speaker.

Its voice replies, "Rainy and cool with a high around 50 degrees."

>> A smart speaker is part of a computer network. It can listen and obey some commands.

You wonder where the cat is. You pick up your phone and tap an icon on the screen. It shows that your furry feline is lounging on his favorite perch by the living room window.

As you leave the house, it automatically locks, but you don't need to take a key. A special camera will recognize your face and unlock the door when you return home.

A microchip is a tiny computer chip that can be placed in a pet to locate it.

"Bye, Zino," you say.

"Goodbye," Zino replies. You know that Zino is only a speaker, but sometimes it seems like there's a real person inside. In fact, your whole house seems to think and feel. What's going on here?

MORE TO EXPLORE

Many science fiction books and movies show **ROBOTS THAT LOOK LIKE PEOPLE** or can move around obstacles and grab objects. A robot, however, can be any machine that does complex tasks on its own.

WHAT MAKES A HOUSE SMART?

Technology is constantly changing our world, so things that seem unusual now may become more common in the future. A smart home is one that contains a computerized communication **network** that can respond to the wants and needs of its residents. The network may control different systems in the home, such as heating, cooling, and lighting. There may also be motion sensors and video and sound monitors that track activity outside and in. This helps keep people safe and secure.

A smart home can have outdoor systems that clear walkways and porches in bad weather.

When someone uses a computer or phone to listen to music, their favorite songs soon appear at the top of the playlist. New singles by the same artist may show up too. How does this happen? A computer collects data, or information, about the tracks the user listens to. This way, the network can get a decent idea of what someone likes.

Using a smart TV, you can shop, browse the internet, catch a movie, and watch your favorite shows.

Similarly, a smart home can learn about the people living there by collecting data. Everything that happens becomes a source of data. If someone always turns up the thermostat a degree or two on chilly mornings, the computer network will learn this, and after a while, it will automatically raise the temperature as needed. A smart home can learn to recognize faces using video monitors. Keypad sensors can record and remember fingerprints.

Voice activation means that someone can speak to a networked device in the house and it will obey commands or answer questions. People talk to the network through a speaker, which they might name. That can make the computer seem a little friendlier and more human, but it's not the only reason. The name acts as a signal. As soon as someone says it, the network becomes active. It turns on and starts to listen.

MORE TO EXPLORE

Voice activation may allow you to **ORDER GROCERIES**, email a friend, or play games. People are developing new uses for voice-activated devices all the time.

PROS AND CONS

As you can see, smart homes have many **conveniences** that help us do things faster and easier. People with physical limitations or disabilities benefit from smart homes too. These homes can help them be independent by automatically doing some housekeeping tasks. Residents can use a voice-activated system to contact doctors, nurses, and other helpers when they are needed.

Smart homes are capable of doing many positive things, so it's difficult to see that they also have some drawbacks. Many people worry about the amount of data that smart homes collect. Thieves called **hackers** can steal data by breaking into a computer network. They may use the information to take money from someone's bank account or to purchase things online and charge them to someone else. They could also control some of the systems in a house. Fortunately, computer experts continue to find ways to make data more secure.

MORE TO EXPLORE

Not all hackers are bad. Companies who develop computer programs use hackers called "WHITE HATS" to locate weaknesses they need to fix.

Sometimes voice-activated devices respond when their owners don't want them to or misunderstand commands. In 2017, a voice-activated speaker overheard a newscaster on TV saying, "I love the little girl saying, 'Alexa ordered me a dollhouse.'" Hearing the statement, speakers at several viewers' homes mistook this as a command and tried to place orders for the dollhouse.

Some users have found ways to avoid this problem. They program their speakers to check with them before sending messages. For instance, the speaker will ask for a special code before making a purchase. If you don't have the code, you'll have to wait for someone who does.

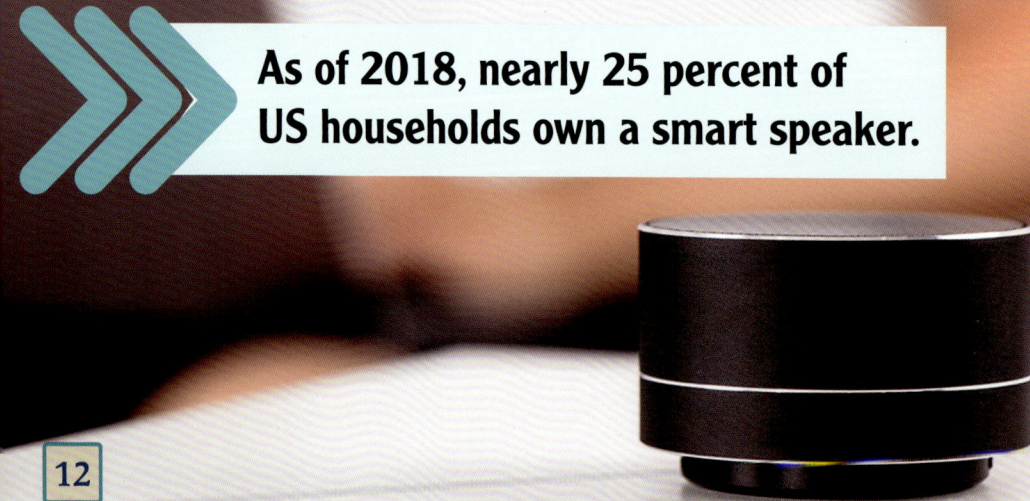

As of 2018, nearly 25 percent of US households own a smart speaker.

Smart technology also costs more than conventional options. Sure, it's convenient to be able to turn on lights or monitor your pet's eating habits with a smartphone, but are those features really worth the money? Consumers have to decide whether to buy high-tech products or more traditional ones.

Do the pros outweigh the cons? That may depend on whether you are an early adopter or a laggard. When it comes to adopting new technologies and innovations, people generally fit into one of five categories, as shown on the next page.

Innovators and early adopters tend to be risk-takers who are typically young and educated or can afford to buy products that may fail. Innovators also often have connections to the product or the scientists and technologists who developed it.

Many people identify as the early majority or late majority; both categories are looking for products proven to work before they spend their hard-earned money on them.

Who are the laggards? Most likely, not you! Laggards tend to be older and less comfortable with technology.

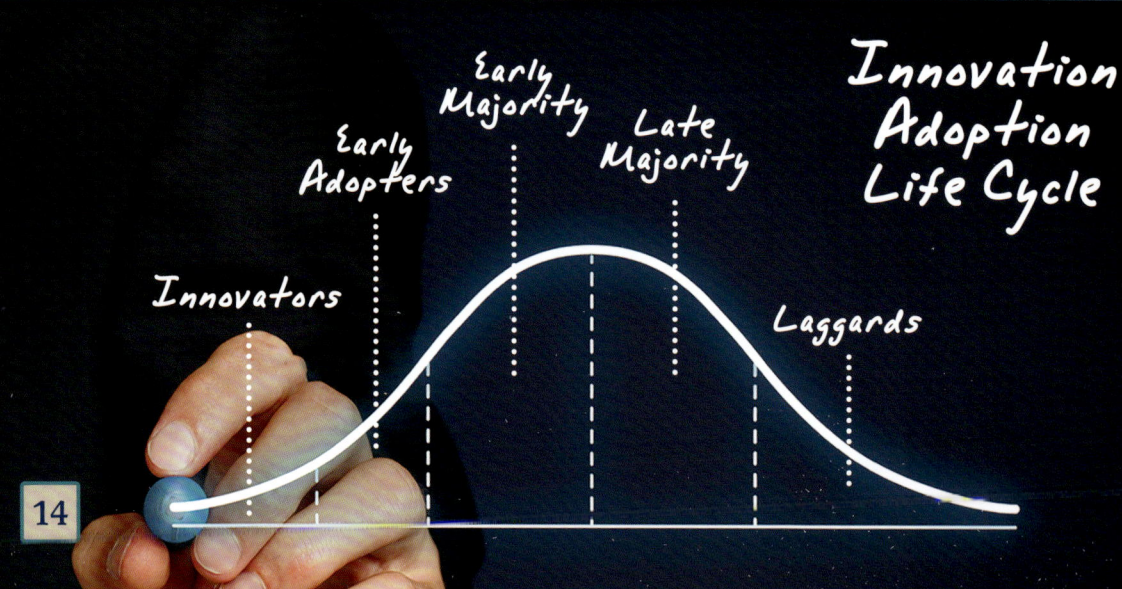

No matter where you or your family fall, smart homes are attracting a lot of interest and will become more and more commonplace in the future. Computers are playing a greater role in our lives every year. Even schools and office buildings are becoming smarter.

MORE TO EXPLORE

In years to come, smart homes, offices, schools, and other buildings may be linked together to **FORM SMART CITIES**. Smart cities can alert drivers to bad weather or traffic emergencies.

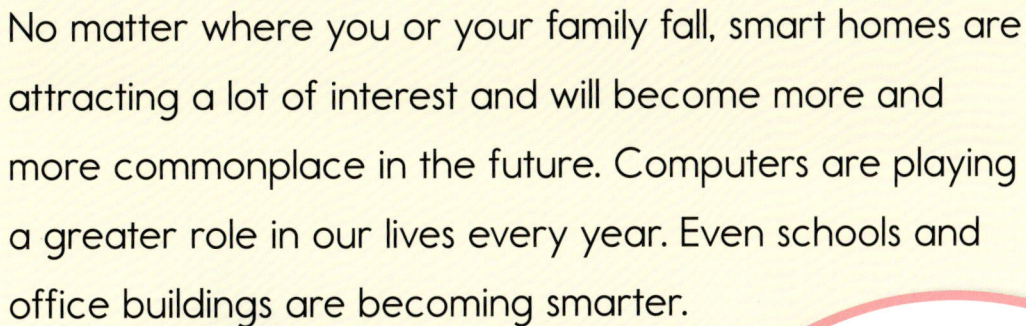

THE FUTURE OF SMART HOUSES

What kind of technology could homes have in the future? Someday residents may be able to change the color of their walls with the touch of a button. Screens in every room could allow people to video chat with others as they move about the house. They may be able to analyze how much energy residents use and become more energy **efficient** on their own. What else can we expect?

Computer-controlled solar panels can help smart homes and cities use the sun's energy wisely.

Home appliances are getting smarter. Kitchens could have refrigerators that keep track of contents and let people know what is available for meals and snacks. They may tell their owners when they need to purchase different foods. New refrigerators might even give recipes for dinner based on their contents.

Robots may take care of more housekeeping chores. If there's a spill, vacuums and electric brooms could automatically clean it up. A robot may even be able to collect, wash, dry, and fold laundry.

Homes may help residents stay healthy. Toothbrushes might be able to measure body temperature and detect fevers. A voice-activated speaker may determine if someone's voice sounds hoarse or raspy, a sign of a sore throat. All this information could be automatically conveyed to health-care providers if necessary.

Most importantly, homes could get even better at learning about the people who live in them. Homes will absorb and analyze data faster than ever. Your television may keep track of sports teams you root for, and the kitchen might even remember your favorite foods. Bedrooms could know their residents' favorite colors and what kind of art they like on the walls.

A smart home might even be able to tell if someone is happy or sad based on the tone of their voice and their movements. If someone is sad, the speaker may start playing some of their favorite music. A home network could also act as a **virtual** counselor, asking about feelings and offering advice. Or it may suggest outside people who can help.

Smart homes are exciting, challenging, and interesting. Making homes better and safer for everyone is a job for the future.

WELCOME TO OUR SMART HOME

In a smart home, many systems and devices are controlled by computers.

SMART APPLIANCES

SMART CAR

SMART FRIDGE

SOLAR PANELS

LIGHT CONTROL

SECURITY CAMERA ACCESS

HOME CONTROL PANEL

SMART TV

DOOR ENTRY FACE RECOGNITION

SMART SPEAKER

GLOSSARY

automatically
happening or working by itself

conveniences
things that simplify a task or an activity

efficient
able to do work without wasting resources

hackers
people who can get information from a computer without the owner's permission

network
a group of computers that are linked together

virtual
being part of the internet or a computer

voice activation
turning on a device using vocal commands

INDEX

automatically 2, 4, 8, 10, 17, 18
cities 15, 16
commands 3, 9, 12, 20
cons 10, 13
conveniences 10
costs 13
early adopter 13–14
efficient 16
hackers 11
home appliances 17
innovation adoption life cycle 14
innovators 14
laggard 13–14
microchip 4
network 3, 6–9, 11, 19
pet 4, 13
pros 10, 13
robots 5, 17
secure 6, 11
smartphone 13
solar panels 16
speaker 3, 5, 9, 12, 18, 19
TV 7, 12
virtual 19
voice activation 9
white hats 11